Neil Armstrong
The First Man on the Moon

Mike Goldsmith

RAINTREE
STECK-VAUGHN
PUBLISHERS

A Harcourt Company

Austin   New York
www.raintreesteckvaughn.com

Titles in this series:
Muhammad Ali: The Greatest
Neil Armstrong: The First Man on the Moon
Fidel Castro: Leader of Cuba's Revolution
Diana: The People's Princess
Anne Frank: Voice of Hope
Bill Gates: Computer Legend
Martin Luther King, Jr: Civil Rights Hero
Nelson Mandela: Father of Freedom
Mother Teresa: Saint of the Slums
Pope John Paul II: Pope for the People
Queen Elizabeth II: Monarch of Our Times
The Queen Mother: Grandmother of a Nation

Published by Raintree Steck-Vaughn Publishers,
an imprint of Steck-Vaughn Company

Library of Congress Cataloging-in-Publication Data

Goldsmith, Mike, Dr.
   Neil Armstrong : The first man on the moon / Mike
Goldsmith.
      p. cm. -- (Famous lives)
   ISBN 0-7398-4431-8
   1. Armstrong, Neil, 1930--Juvenile literature. 2. Astronauts--
United States--Biography. 3. Project Apollo (U.S.)--Juvenile
literature. [1. Armstrong, Neil, 1930- 2. Astronauts. 3. Space
flight to the moon. 4. Project Apollo (U.S.)] I. Title. II.
Famous lives (Austin, Tex.)

TL789.85.A75 G65 2001
629.45'0092--dc21
[B]
                                              2001019209
Printed in Italy. Bound in the United States.

1 2 3 4 5 6 7 8 9 0 LB 06 05 04 03 02

Cover: Neil Armstrong in a space suit ready for launch.
Title page: Earth as seen from near the Moon's surface

Picture Acknowledgments
The publisher would like to thank the following for giving
permission to use their pictures: Corbis (cover), 35, 43 (top and
bottom), 45; Genesis (title), 7, 20, 22, 27, 28, 32, 36, 39, 40,
41, 42; Ohio Historical Society 8, 9(top), 10, 17; Photri 4, 5,
9(bottom), 13, 16, 18, 19, 21, 23 (top and bottom), 24, 25, 26,
29, 30, 31, 33, 34, 37, 38, 44; Popperfoto 6, 11, 12, 15;
T.R.H./US Navy 14.

While every effort has been made to secure permission, in some
cases it has proved impossible to trace copyright holders.

Contents

First Landing

The Eagle *leaving the Moon's surface. Earth can be seen over the lunar horizon.*

It was July 20, 1969, and a spacecraft was approaching the surface of the Moon. It was called the *Eagle*, and it was one of the last remaining pieces of an enormous machine that had blasted off from Earth three days before. For most of its journey, the ship had coasted through space under the influence of the gravity of the Moon and Earth. The final approach to the Moon had been mostly computer-controlled, but now it was time for Neil Armstrong to take over.

Neil Armstrong in the Eagle's cabin, having just returned from his exploration of the Moon's surface.

As the Moon's surface loomed closer, Armstrong saw that they were heading straight towards a field of huge boulders. If they tried to land there, the ship would be destroyed and Armstrong and his fellow astronaut, Buzz Aldrin, would die on the Moon.

Armstrong calmly steered the spaceship away from the boulders, but wherever he looked there were no places to land safely. Finally, with only about a minute's worth of fuel remaining, he saw a clear space ahead. Touchdown was so gentle neither astronaut felt it. Armstrong switched off the engine, and he and Aldrin looked out through the windows at the airless, desertlike landscape of the Moon.

Destination Moon

People have always been fascinated by the Moon. The first calendars were based on its regular cycle of changes from the new to full Moon, and there many legends about the Moon and what the markings on its face are. When the telescope was invented in the early 17th century, the Moon's surface could finally be seen. It looked like the Earth—in some ways. Mountains and plains could be clearly seen, but was there water, air, or even life?

"The Moon certainly does not possess a smooth and polished surface, but one rough and uneven, and, just like the face of Earth itself, is everywhere full of vast mountains, deep valleys, and sinuosities [many forms]."
Galileo's report of his first telescopic observations of the Moon in 1610.

The Italian scientist Galileo Galilei with one of the many telescopes he made.

Although the first spacecraft from Earth only reached the Moon in the 1960s, there have been stories about traveling to the Moon for more than a thousand years. It was said that the fictional travelers were able to get there by using strange sources of power—including the wind, sunlight, and even migrating geese! There were scientific reasons to go to the Moon (like finding out where it came from) but the main fascination was just to reach a different world that seemed so very close.

The Moon, nearly full, photographed from Apollo II.

Birth of an Astronaut

"I must have been a staunch aviation fan before I was even conscious of it."
Neil Armstrong.

Neil Armstrong was born on August 5, 1930, in Wapakoneta, Ohio. Neil's dad was an accountant who often had to move the family from one town to another to find work—so Armstrong had many homes in different parts of Ohio. One of these was in Cleveland, which had its own airport. When he was only two, his parents took him there to watch the planes. He soon had a fascination with flying that was to last for the rest of his life and eventually take him to the Moon.

In 1936, the family were living in the town of Warren. In those days, very few civilians—and almost no six-year-olds, had the opportunity to fly. Armstrong would soon get his chance. One Sunday he and his father sneaked off from church to pay a visit to Warren Airport to see a plane nicknamed "the Tin Goose." The pilot offered them a ride and Armstrong flew for the first time. He loved it, and soon he was spending all his free time building model planes.

Neil Armstrong was already fascinated by planes, even at the early age of two.

Right **Neil Armstrong, age four, with his sister June.**

Below **A Ford Trimotor—the type of plane in which Neil learned to fly.**

War in Europe

Armstrong's fascination with planes continued, and he was determined to fly. To do that, he had to take flying lessons, and to do *that* he needed to earn some money. Armstrong found a job helping out at a shop where he had to work for over 22 hours to afford a single lesson! But he was dedicated and kept up the work and the flying lessons.

After two years of lessons, and on his 16th birthday, he got his license. At school Armstrong worked hard, but he didn't spend all of his time studying: he started a jazz band, in which he played the horn, and he also joined the Boy Scouts.

"He always had a goal to work on."
Armstrong's high school teacher.

A young Neil Armstrong in his band uniform.

In 1939, World War II had broken out in Europe, and the United States was drawn into the conflict in 1941. At the time it had had little effect on Armstrong, but he would soon become personally involved in the war.

A World War II American pilot poses on a Mustang in 1944.

Call-up for Korea

When he left high school, Armstrong was determined to go to college to follow up on his interest in aeronautics—the science of flight. The only problem was money—until the Navy offered him a scholarship. In 1947, he began studying aeronautical engineering at Purdue University in Indiana, where he also joined the Naval Air Cadet program.

"I was in the basement getting out quart jars of canned fruit to bake some pies. He called so loudly, 'Mom, Mom,' that he scared me to death. I dropped a jar of blackberries on my big toe. I must have broken the toe—it was black and blue for weeks. But that was such a great day."
Viola, Armstrong's mother, remembering the day Armstrong got his scholarship.

A pilot of a B-29 plane prepares his crew for a bombing raid on Korea.

Soon after Armstrong started his college course, a war started in Korea. The United States joined the South Koreans to fight against the Chinese-supported North Koreans. The Navy selected young men with the skills they needed, and Armstrong was an obvious choice. In 1950 he was called up to fight. After some rapid training in Florida, he was assigned to an aircraft carrier—the USS *Essex*—as a fighter pilot. His dreams of being a pilot had now become true.

The USS Essex, the aircraft carrier from which Neil Armstrong flew over Korea.

Years of Danger

Armstrong flew 78 missions in the Korean War. Though he had trained in air-to-air combat, his job was to fly into enemy territory to damage or destroy bridges and other targets. It was dangerous work. To keep under cover, Armstrong had to fly very low, which required split-second reactions and excellent flying skills.

"I didn't want to be responsible for anyone else." Armstrong's reason on why he preferred to be a solo pilot.

*The type of plane called **The Panther** was the model that Neil Armstrong flew in Korea.*

In September 1950, under cover of massive air strikes, American troops landed in Korea.

It wasn't just natural hazards that Armstrong had to face. Once, the tip of one of his plane's wings was sheared off by a cable placed in his path as a booby trap. It took all of Armstrong's flying skills to coax the damaged plane out of enemy territory, so that he could parachute to safety. On another occasion he managed to fly a badly damaged plane to a safe landing. Many of Armstrong's fellow pilots died in Korea.

In 1952, when his tour of duty was over, Armstrong returned to the United States to complete his college degree. He later received three medals for his courage in the war.

Falling in Love

Neil Armstrong at Edwards Air Force Base, during training in 1956.

> **"He flies an airplane like he's wearing it."**
> One of Armstrong's fellow pilots at Edwards Air Force Base.

In 1955, Armstrong finished his college degree course and, still fascinated by flying, took a job as a pilot. But it was no ordinary flying job. He was posted to Edwards Air Force Base in California, where a government organization called the National Advisory Committee for Aeronautics (NACA) was developing a new generation of rocket-propelled planes, which flew higher and faster than anything else in the sky.

In 1956, Armstrong married Jan Shearon. They had met four years earlier, at Purdue University, but it was a long time before they went out together and even longer before they married. As Jan said, "Neil isn't one to rush anything."

Armstrong had always been a very private person so he and Jan made a home in a forest cabin in the San Gabriel mountains in California, rather than in the town of Lancaster where most of his colleagues at the air force base lived.

Neil Armstrong proposed to Jan Shearon in the summer of 1955, when she was working at a summer camp in Wisconsin. They were married the following January.

Moon Race

In 1957, the Armstrongs had their first son, Eric. The Space Age was born the same year when the Soviet Union launched Sputnik, the first satellite, into space. The rest of the world was amazed. The U.S. government and public were concerned that the Soviet Union had beaten them into space. They were even more worried when the Soviets put the first animal and the first human into space. The animal was a dog called Laika, and the man was the cosmonaut, Yuri Gagarin.

Yuri Gagarin, who at 27 became the first man in space on April 12, 1961.

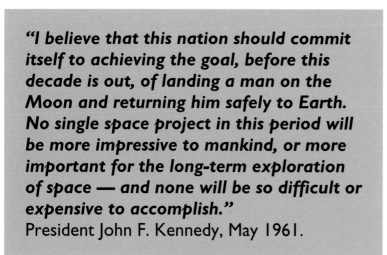

The United States was determined to catch up, and in 1961 President John F. Kennedy announced his determination to land an American on the Moon by 1970. It was a bold decision. At the time there had been only one American astronaut, Alan Shepard, and he had spent just 15 minutes in space.

Alan Shepard, who became the first American in space on May 5, 1961.

"I believe that this nation should commit itself to achieving the goal, before this decade is out, of landing a man on the Moon and returning him safely to Earth. No single space project in this period will be more impressive to mankind, or more important for the long-term exploration of space — and none will be so difficult or expensive to accomplish."
President John F. Kennedy, May 1961.

The Edge of Space

Neil Armstrong in the cockpit of the X-15 experimental rocket-plane.

In 1958, NACA changed its name to NASA, the National Aeronautics and Space Administration, and selected the first seven astronauts to pilot a series of one-man Mercury capsules into space. Neil Armstrong was not among those who volunteered. For one thing, astronauts then were more like passengers than pilots. Pilots of rocket-planes, on the other hand, had to use all their skills on their high-speed, high-altitude missions. On one flight in 1962, in the X-15 rocket-plane, Armstrong reached 39 miles (63 km) above Earth—the edge of space. In total, Armstrong flew over a thousand hours in experimental rocket-planes like the X-15.

"Space is the frontier, and that's where I intend to go."
Neil Armstrong in 1962.

Armstrong was to have piloted a new type of rocket-plane, bigger and faster than the X-15, called the Dyna-Soar. However, while this new rocket-plane was still on the drawing board, priorities changed. NASA's main goal was no longer to send rocket-planes to the limits of the atmosphere—it was to send a man to the Moon. When the next opportunity came, Armstrong volunteered as an astronaut—and was accepted.

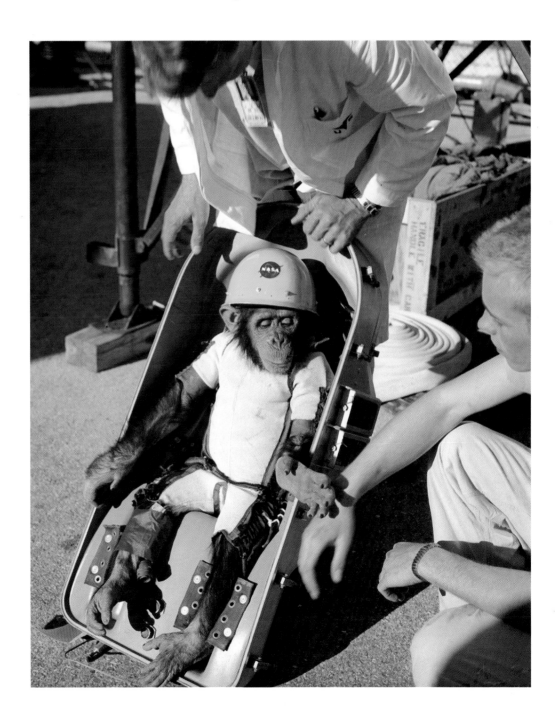

Before risking the lives of human astronauts, NASA sent animals into space, including a chimp called Ham. Like most test animals, Ham returned safely to Earth.

Training for Space

There was a great deal to learn about space travel before it would be possible to travel to the Moon, so NASA designed and launched a series of spacecraft to prepare for lunar missions. The first craft had no people on board, but later models carried two astronauts. These were called Gemini spaceships because Gemini is a constellation of stars that resembles a pair of twins.

Armstrong's first trip into space was as a Gemini astronaut, but before he could go he had to take some tough training. He was spun at high speeds to simulate the accelerations he'd experience. He spent hours in a model of a Gemini cockpit, learning exactly how to fly it, and how to cope if things went wrong. He was even taught how to survive if his spacecraft came down in a jungle.

The crew of Gemini 8— Dave Scott and Neil Armstrong—pose for publicity photographs before their mission.

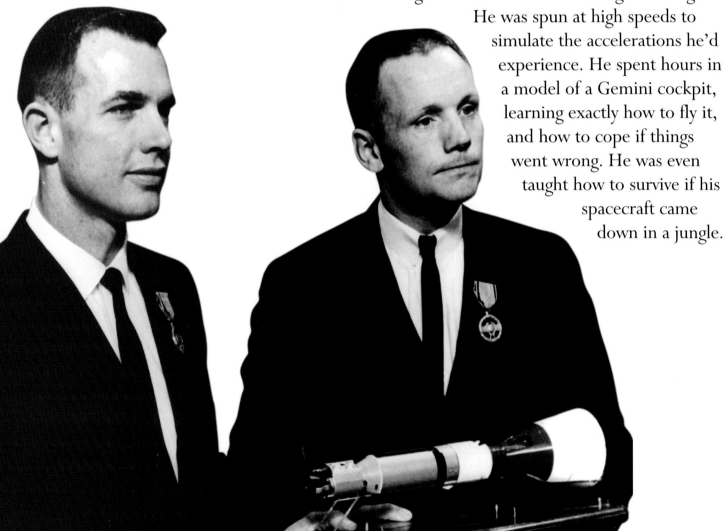

Armstrong was a member of the backup crew for *Gemini 5*, and then he was selected to be the commander of the *Gemini 8* mission in 1966. He had his chance to be a space-traveler at last. He didn't suspect how close he was about to come to death in space.

> **"Their training had been fantastically intense; it had little to do with the public image of an astronaut jogging on the beach."**
> From Gene Farmer's *First on the Moon*.

Above **Dave Scott and Neil Armstrong (background) learn to cope with the weightless conditions they would undergo in space.**

Left **Neil Armstrong, takes his temperature as part of a medical check-up in preparation for his Gemini mission.**

23

Danger in Space

Armstrong's first space mission took place in 1966 with fellow astronaut Dave Scott. Their job was to join together two spacecrafts in space, an operation known as docking, which had never been attempted before.

The launch of **Gemini 8.**

The experiment worked perfectly—Armstrong and Scott gently guided their spaceship to a special satellite, and then locked the two together. But then, when they were 186 miles (300 km) above the Earth, the linked spaceships began to tumble. Whatever Armstrong did had no effect, and the spin got faster and faster. He successfully separated the two ships, but things only got worse, and soon *Gemini 8* was turning over every few seconds, and still getting faster. If Armstrong didn't do something soon, the ship would disintegrate and they would die in space.

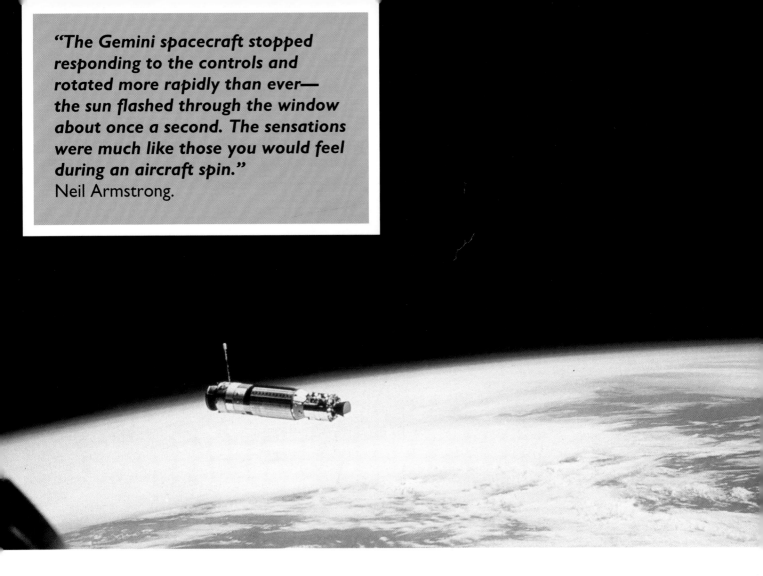

> *"The Gemini spacecraft stopped responding to the controls and rotated more rapidly than ever—the sun flashed through the window about once a second. The sensations were much like those you would feel during an aircraft spin."*
> Neil Armstrong.

The Agena spaceship photographed from **Gemini 8** *before docking.*

Armstrong had one more try—he switched off all the steering thrusters, and switched on the re-entry thrusters in the spaceship's nose. It worked! Gradually the spinning slowed and stopped.

Back on Earth the cause of the problem was soon discovered —one of the steering thrusters had jammed in the "on" position. If Armstrong hadn't switched them all off, his first space mission would have been his last.

25

A Tragic Beginning

None of the Apollo crews died in space, but the astronauts who were to have been on the first Apollo mission died tragically in 1967 when their capsule was being tested. They were killed by a fire that swept through the capsule in seconds. The astronauts were trapped because the hatch, which opened inward, was held shut by the high air pressure in the capsule.

"If we die, we want people to accept it. We're in a risky business."
Astronaut Gus Grissom, a few weeks before his death in *Apollo I*.

The scorched remains of the Apollo I capsule.

Armstrong himself had another narrow escape in 1968.
This time he was flying the Lunar Landing Research Vehicle,
an aircraft that was built to behave like the Lunar Module that
would eventually land on the Moon. When Armstrong was
flying the machine for the twenty-first time, and it was over
98 feet (30 m) in the air, it started to tilt sharply to one side.
As the aircraft went completely out of control, Armstrong
managed to eject himself just in time. While it crashed to the
ground and exploded, he landed safely by parachute.

**Neil Armstrong in a
simulation of the cabin
of an Apollo capsule.**

The Moon-Landing Mission

The Apollo astronauts were assigned to their missions in strict rotation. Armstrong's first assignment was as a backup crewmember of the *Apollo 8* mission, in case any of the astronauts needed to be replaced. Then, in January 1969, he was assigned to *Apollo 11*, the Moon-landing mission. Soon after, Armstrong was told that he would be the first person to step onto the Moon. His fellow crew members were Buzz Aldrin and Mike Collins. Aldrin and Armstrong were to travel down to the Moon's surface in the Lunar Module while Collins was to remain in orbit around the Moon in the Command Module.

The crew of Apollo 11: Neil Armstrong, Mike Collins, and Buzz Aldrin.

The next few months were a period of intense activity for Armstrong, Aldrin, and Collins, as they spent up to 14 hours a day practicing on "simulators"—imitations of the cockpits of the Lunar and Command Modules. There were also hours of lectures on rocketry, astronomy, and geology, and many medical tests.

"What I really want to be, in all honesty, is the first man back from the Moon."

Neil Armstrong and Buzz Aldrin training for their exploration of the Moon.

The Moonship

The spacecraft that would take Armstrong and his crew to the Moon was enormous, very powerful, and extremely complicated—and it also had to be very reliable.

The spaceship was composed of eight main parts, seven of which would be discarded at different stages of the journey. This made the spaceship lighter and therefore meant it needed less fuel.

The three stages of the *Saturn 5* rocket were mainly fuel tanks and rocket motors, used to drive the spaceship into space and start it on its journey to the Moon.

A diagram of the Saturn 5 rocket.

escape tower

service module

command module

lunar module (in two parts)

third stage

second stage

first stage

exhaust burners

APOLLO
LUNAR LANDING MISSION PROFILE

CSM-LM RENDEZVOUS

TRANSEARTH INJECTION

LIFTOFF

CSM TRANSEARTH TRAJECTORY (55-60 HRS.)

INSERTION

LM ASCENT

LUNAR ORBIT RETURN

CM WATER RECOVERY (PACIFIC)

LAUNCH

60 N. M. ORBIT

LUN. SURF. EXPERIMENTS

TOUCHDOWN

LUNAR ORBIT INSERTION

BEGIN LUNAR ORBIT

DESCENT

TRANSLUNAR INJECTION

CSM TRANSLUNAR TRAJECTORY (65-75 HRS.)

NASA HQ FP69-1654I 6-11-69

When the escape tower and the three stages had been used and jettisoned, the three Apollo Modules would be left. The Lunar Module was the only half that would actually land on the Moon, and only the upper part of it would take off again. The job of the Service Module was to complete the journey to the Moon, orbit it, and then take the astronauts most of the way home. It would also finally be discarded and only the Command Module would return to Earth.

*A diagram of the **Apollo 11** lunar-landing mission.*

> *"It is a monster, that rocket. It is not a dead animal; it has a life of its own."*
> Apollo rocket scientist Gunter Wendt.

Escape from Earth

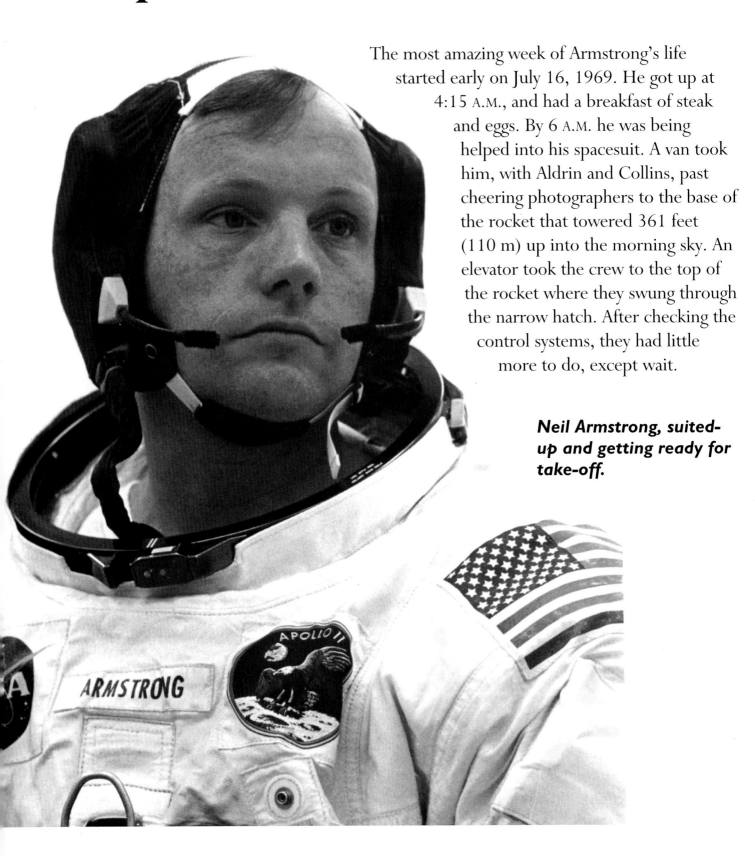

The most amazing week of Armstrong's life started early on July 16, 1969. He got up at 4:15 A.M., and had a breakfast of steak and eggs. By 6 A.M. he was being helped into his spacesuit. A van took him, with Aldrin and Collins, past cheering photographers to the base of the rocket that towered 361 feet (110 m) up into the morning sky. An elevator took the crew to the top of the rocket where they swung through the narrow hatch. After checking the control systems, they had little more to do, except wait.

Neil Armstrong, suited-up and getting ready for take-off.

The launch of Apollo 11, with Neil Armstrong, Mike Collins, and Buzz Aldrin on board.

At 9:32 A.M. the Saturn's rockets fired with a deafening roar and a gush of bright yellow flames. Very slowly it started to lift from the launch pad, then gathered speed rapidly and raced up into the sky. Forty seconds later it was going faster than sound, and only 2.5 minutes after launch the huge first stage fell back to Earth. It was followed just six minutes later by the second stage. The rocket's third stage took it right out of the atmosphere and sent it into orbit 114 miles (184 km) above Earth. The first part of the journey to the Moon had gone perfectly.

"At my window right now I can observe the entire continent of North America."
Neil Armstrong

A Voyage Through Space

The spaceship didn't orbit Earth for long. Less than three hours after take-off the third-stage rocket fired again, launching the ship away from Earth and towards the Moon at 24,109 miles (38,800 km) per hour.

Except when the rockets were fired, everything on the ship was weightless: nothing fell, and everything that wasn't fixed down drifted around the cabin, including the crew.

The Command and Service Module and Lunar Module abandon the third stage of the Saturn 5.

After a tricky maneuver to realign the Apollo Modules, the third stage was abandoned and the rest of the ship continued on toward the Moon, rotating slowly to avoid any parts being in the Sun too long and overheating.

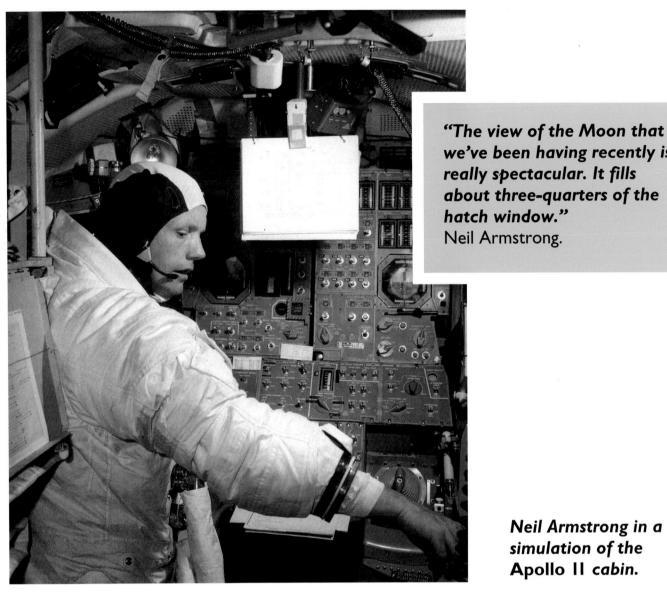

"The view of the Moon that we've been having recently is really spectacular. It fills about three-quarters of the hatch window."
Neil Armstrong.

Neil Armstrong in a simulation of the Apollo 11 cabin.

On board, the crew were in constant contact with Mission Control on Earth. Cameras on board the ship could even transmit pictures back to Earth. All over the world, hundreds of millions of people looked in on Armstrong and his crew.

Apart from a single correction to the flight-path, the rocket traveled unpowered toward the Moon. It gradually slowed down as Earth's gravity tried to pull it back—until 62.5 hours after launch, when the Moon's gravity took over. Then the spaceship started to pick up speed again for the final stage of its journey.

Journey's End

> "It's a strange, eerie sensation to fly a lunar landing trajectory —not difficult, but somewhat complex and unforgiving."
> Neil Armstrong

Before Neil Armstrong's lunar mission, the Apollo 10 astronauts carried out a test-flight and achieved lunar orbit in the Charlie Brown Command and Service Module, just as Columbia would.

Three days after launch, the spaceship's rocket motor fired briefly, changing its course to bring it into orbit around the Moon. After a second course adjustment, the ship was only 62 miles (100 km) above the lunar surface. Looking out, Armstrong and the others could see ranges of ancient mountains, great plains of dust and, wherever they looked, craters of all sizes.

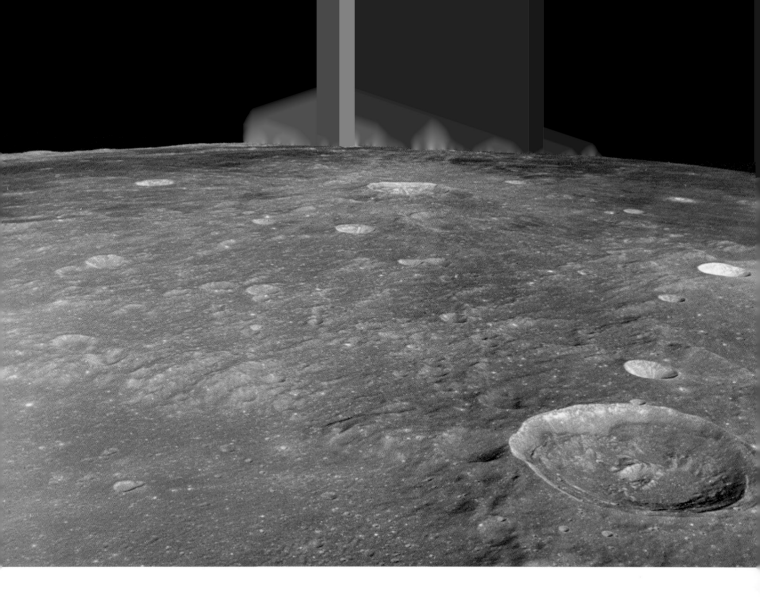

The next stage was to prepare the Lunar Module—now called the *Eagle*—for its landing on the Moon. It was filled with oxygen, power was transferred to it, and its landing legs were swung into position. Finally, Armstrong and Aldrin sealed the hatch and powerful springs separated it from the Command and Service Module (now called the *Columbia*). The rocket motors of the *Eagle* fired and the little spaceship began its descent.

The flight wasn't as straightforward as Armstrong had hoped. The computer showed that they were off course slightly, and as they neared the surface, several alarm signals sounded—there was too much information for the computer to cope with. Undaunted, Armstrong guided the *Eagle* down to a perfect landing. They were on the Moon.

The Apollo 12 Lunar Module Intrepid approaches the Moon's surface, carrying a new crew of astronauts for the second manned landing.

A New World

Neil Armstrong took this picture of Buzz Aldrin about to step onto the Moon.

The plan was that the astronauts would rest before leaving the *Eagle*, but they were too excited to wait. They sealed their spacesuits, allowed the oxygen in the cabin to escape, and opened the hatch. Then Armstrong crawled carefully out of the *Eagle*, down the ladder, and stepped onto the surface of the Moon. A few minutes later, Aldrin joined him.

"That's one small step for man, one giant leap for mankind." Armstrong's first words from the lunar surface.

The men had less than two hours to explore their new world—and they had plenty to do. They set up a lunar earthquake detector and a special type of mirror. A laser beam from Earth would be directed at it, to measure the distance to the Moon to within six centimeters. They experimented with different ways of moving around in the low-gravity environment of the Moon. They also collected rocks and dust, set up an American flag, and spoke to Richard Nixon, the U.S. President. Armstrong caught sight of some strange metallic globules and some clear crystals in the dust, but he had no chance to collect them. All this was watched on television by more than one-fifth of the world's population: over five hundred million people.

Buzz Aldrin exploring the Moon. The Eagle *is in the background.*

Return to Earth

*A painting shows the upper half of the **Eagle** taking the astronauts back to **Columbia**. The lower half was left on the Moon.*

All too soon, it was time to return to the *Eagle*. Armstrong and Aldrin climbed back up the ladder, filled the cabin with oxygen again, and tried to sleep for a while. Then, 12 hours after landing on the Moon, it was time to leave. Armstrong triggered the rocket motor, and the *Eagle* separated into two halves—one half remained on the Moon while the other sped back up toward *Columbia* where Mike Collins was patiently waiting for them.

Three hours later, *Eagle* and *Columbia* docked—the maneuver that Armstrong had practiced on his first, near-fatal space trip three years earlier. The astronauts were reunited, and the *Eagle* was allowed to fall back to the Moon, while *Columbia* started its long journey home. On the way, the astronauts were puzzled by some strange flashing lights in their eyes. Later, scientists decided the flashes were due to cosmic rays.

The Earth as seen from the Moon. For the astronauts, it was still a long way home.

"The responsibility for this flight lies first with history and with the giants of science who have preceded this effort." Neil Armstrong, on the journey back to Earth.

Splashdown

Navy divers met the Command Module after it splashed down into the Pacific Ocean. When leaving the Command Module, the astronauts wore special clothing to protect other people from any germs they might have picked up on the Moon.

After three days in space, *Columbia* was approaching Earth again. After its rocket fired once more to correct their course, there was no further use for the Service Module, so it was jettisoned. Now only the flying-saucer-shaped Command Module was left to return the astronauts home. But if it flew directly into Earth's atmosphere, it would burn up. So it flew a zigzagging path, swooping through the atmosphere briefly to lose speed before skimming out again to cool down. When it had done this several times, it was moving slowly enough to survive the last stage of the journey. It raced down through the atmosphere, glowing red-hot. The astronauts had to endure the effects of enormous changes in speed—after being weightless for so long, they had to cope briefly with weighing six times more than usual.

Neil Armstrong, Mike Collins, and Buzz Aldrin (left to right) look out through the window of their quarantine capsule.

Finally, the parachutes opened and the Command Module splashed down into the Pacific Ocean. An hour later a helicopter was taking it back to the United States with Armstrong, Aldrin, and Collins still on board. The first men on the Moon had returned to Earth.

"Neil, Buzz, and Mike, I want you to know that I think I'm the luckiest man in the world, and I say this not only because I have the honor to be President of the United States, but particularly because I have the privilege of speaking for so many in welcoming you back to Earth."
President Nixon, July 1969

Jan Armstrong, with sons Eric and Mark, facing an army of reporters.

After Apollo

For 21 days after returning to Earth the astronauts had to stay in quarantine in case they had picked up any alien germs on their trip. Armstrong, Aldrin, and Collins then went on a world tour, visiting 28 cities in 25 countries. Among the honors Armstrong received, one of the strangest was being sculpted in 880 pounds (400 kg) of butter by the Ohio Dairymen's Association.

Armstrong has never been keen on publicity, and after the tour he did his best to avoid it. He rarely grants interviews and only occasionally makes public appearances. After working for a few more years at NASA, he became Professor of Aerospace Engineering at the University of Cincinnati, Ohio, and bought a farm nearby. He left the university in 1979.

New York City welcomes the returning astronauts with a ticker-tape parade.

Neil Armstrong at a ceremony in 1999 to collect the Langley Medal, which is awarded for the application of science to flight.

In 1986, the U.S. Space Shuttle *Challenger* exploded soon after take-off, killing its crew of seven. Armstrong was vice-chairman of the inquiry to establish the cause of the disaster.

Armstrong is now chairman of AIL Systems Incorporated, a New York electronics company, but he still lives in Ohio. The footprints he left on the Moon will be there for centuries.

"I don't want to be a living memorial."
Neil Armstrong

Glossary

Aeronautics (air-uh-NAW-tiks) The science of air flight.

Air-to-air combat When an aircraft fights other aircraft in the air.

Assigned To be given a task or a position.

Atmosphere (AT-muhss-feer) The layer of gases which surrounds Earth.

Booby trap A harmless-looking device which can kill or injure anyone who touches it.

Boulders (BOHL-durs) Large rocks.

Civilians (si-VIL-yuhns) People who aren't in the armed services or the police force.

Coasted Traveled without engine power.

Constellation (kon-stuh-LAY-shuhn) A pattern of stars.

Cosmic rays Mysterious high-energy radiation found in space.

Cosmonaut A Soviet astronaut.

Crater A deep hollow with a raised rim.

Discard To get rid of something when you have finished with it.

Disintegrate To crumble away into fragments.

Docking Joining two spacecraft together in space.

Eject To quickly leave an aircraft in an emergency.

Gravity The force that keeps people on the ground, makes stones fall to Earth and keeps the Moon orbiting Earth and the planets orbiting the Sun.

Hatch A door or opening in an aircraft or a spacecraft.

Jettison To get rid of something no longer needed.

Lunar (LOO-nur) To do with the Moon.

Mission Control The group of people on the Earth who monitor space missions and remain in contact with the astronauts throughout.

Orbit The path of one object around another in space.

Quarantine (KWOR-uhn-teen) A set period of time when a person is kept away from other human beings in case they have been infected with diseases that can be passed on.

Rotation When people in a team take it in turns to be on duty on a regular basis.

Satellite (SAT-uh-lite) A natural or artificial object orbiting a planet.

Simulate To imitate conditions so closely that the simulation seems like the real thing.

Sinuosities (sin-yuh-WAH-suh-tees) Winding shapes.

Space shuttle A reusable spacecraft, which is launched with a rocket but lands like a plane.

Thruster A small rocket used to steer a spacecraft.

Touchdown The moment an aircraft or spacecraft lands.

Volunteer To offer your services of your own free will.

Date Chart

August 5, 1930 Neil Alden Armstrong born in Wapakoneta, Ohio.
1936 First flight.
1946 Obtains pilot's license.
1947 Starts studying aeronautical engineering at Purdue University.
1950 Called up to fight in Korea.
1952 Returns to Purdue University.
1955 Graduates with degree in Aeronautical Engineering.
1956 Marries Jan Shearon.
1957 Son Eric born. Sputnik launched. Armstrong becomes a test pilot.

1959 Daughter Karen born.
1961 President Kennedy announces plan to land a person on the Moon by 1970.
1962 Becomes an astronaut trainee.
1963 Son Mark born.
1966 First space mission, as commander of *Gemini 8*.
1969 Becomes first human on Moon.
1971–1979 Works as Professor of Aerospace Engineering at University of Cincinnati.
1986 Vice-chairman of *Challenger* inquiry.

Further Information

Books to Read
Bond, Peter. *A Photographic Journey through the Universe*. Dorling-Kindersley, 1999.
Bredeson, Carmen. *Neil Armstrong*. Enslow, 1998.
Chaikin, Andrew. *A Man on the Moon*. Penguin, 1995.
Connolly, Sean. *Neil Armstrong*. Heinemann, 1999.
Scott, Carol and Sarah Ashun. *Eyewitness: Space Exploration*. Dorling-Kindersley, 1998.
Twist, Clint. *Gagarin and Armstrong: The First Steps in Space*. Evans Brothers, 1996.

Websites
NASA's official website dedicated to aeronautics and space research: www.nasa.gov
The following websites are designed for children in grades 4–8 who want to learn more about Space Science and Technology.
NASA KIDS
http://kids.msfc.nasa.gov
The Space Place
http://spaceplace.jpl.nasa.gov
NASA Quest—Brings NASA people and science to K–12 classrooms through the Internet http://quest.arc.nasa.gov/space/index.html
Popularization of space and all space-related subjects: www.space.com

Index

Page numbers in **bold** mean there is a picture on the page.